Backhoes

Mari Bolte

CREATIVE EDUCATION • CREATIVE PAPERBACKS

seedlings

Published by Creative Education and Creative Paperbacks
P.O. Box 227, Mankato, Minnesota 56002
Creative Education and Creative Paperbacks
are imprints of The Creative Company
www.thecreativecompany.us

Design by Ellen Huber
Production by Alison Derry
Art direction by Tom Morgan

Photographs by Alamy (GerryRousseau), Dreamstime (Dmitry Kalinovsky, Orangeline, Stefan11), Getty (Cretì Stefano, Steve Cicero), iStock (ewg3D), Shutterstock (6493866629, Alekcey, ChiccoDodiFC, Digital Genetics, Fahroni, Milos Stojanovic, Nerthuz, Sergei Dvornikov, TheHighestQualityImages, Th0ngsuk sasim, Valentin Valkov)

Library of Congress Cataloging-in-Publication Data
Names: Bolte, Mari, author.
Title: Backhoes / Mari Bolte.
Other titles: Seedlings (Creative Education, Inc. (Mankato, Minn.))
Description: Mankato, Minnesota : Creative Education and Creative Paperbacks, [2024] | Series: Seedlings | Includes bibliographical references and index. | Audience: Ages 4–7 | Audience: Grades K–1 | Summary: "An early elementary-level STEM introduction to the backhoe, covering how the construction vehicle looks and works. Includes a glossary, further resources, and a labeled image guide to the digging machine's major parts"— Provided by publisher.
Identifiers: LCCN 2023012520 (print) | LCCN 2023012521 (ebook) | ISBN 9781640267565 (library binding) | ISBN 9781682773062 (paperback) | ISBN 9781640009264 (pdf)
Subjects: LCSH: Backhoes—Juvenile literature. | Excavating machinery—Juvenile literature. | Construction equipment—Juvenile literature. | CYAC: Backhoes. | Excavating machinery. | Construction equipment. | LCGFT: Instructional and educational works.
Classification: LCC TA735 .B585 2024 (print) | LCC TA735 (ebook) | DDC 621.8/65—dc23/eng/20230329
LC record available at https://lccn.loc.gov/2023012520
LC ebook record available at https://lccn.loc.gov/2023012521

Printed in China

TABLE OF CONTENTS

Time to dig!

Backhoes are small diggers. They make holes in the ground.

They move things
from place to place.

Backhoes have two buckets. Each bucket has its own job.

SURFWOOD
JCB

The bucket on the front is the loader. It scoops up dirt, rocks, and snow. Two **arms** lift the load into a dump truck.

The bucket on the back digs holes. A strong arm moves it. Teeth help the bucket break through the hard ground.

Backhoes move piles of rocks. They pull up tree stumps. They also dig **ditches**.

Backhoes are useful tools. They are two machines in one!

All done digging!

Picture a Backhoe

cab
teeth
digger bucket
wheel

Words to Know

arm: a long part that can reach and move

ditch: a long, deep cut in the ground

teeth: sharp parts of a tool that help it grab or dig

Read More

Hughes, Susan. *Hooray for Trucks!* Berkeley, Calif.: Owlkids Books, 2022.

Humphrey, Natalie. *Great Giant Cranes.* New York: Gareth Stevens Publishing, 2023.

Websites

Construction Fun Facts
https://www.cannykids.com/construction-fun-facts
Find out more about different types of construction vehicles.

Construction Trucks!
https://supersimple.com/caities-classroom/construction-trucks
Facts, activities, and construction truck projects for all ages.

Index